# The Blue Lobster

# The Blue Lobster
## *a life cycle*

Carol and Donald Carrick

The Dial Press · New York

Library of Congress Cataloging in Publication Data
Carrick, Carol. The blue lobster.
1. Lobsters—Juvenile literature. [1. Lobsters]
I. Carrick, Donald, illus. II. Title.
QL444.M33C37     595'.3841     74-18594
ISBN 0-8037-4482-X
ISBN 0-8037-4483-8 lib. bdg.

For Ted, who dreamed of the blue lobster

With special thanks for his help to John T. Hughes,
Director, State Lobster Hatchery and Research
Station at Martha's Vineyard, Massachusetts

The last rays of the sun melted across the floor of the summer sea. They glowed on a rock ledge where a mother lobster had taken shelter.

She glided out on the tips of her slender legs, her body heavy with the thousands of ripe eggs she had carried for a year and a half.

The eggs were cemented with a sticky fluid to the hairs on her swimmerets, a double row of swimming paddles under her body. Even when she rested, the swimmerets beat back and forth to keep the eggs fresh and clean. Whenever she sensed danger, her armored tail curved round her eggs to shield them.

That evening they were ready to hatch. She raised her tail and violently shook her swimmerets till hundreds of baby lobsters streamed away from her and rose toward the fading light. On the next few nights she would release hundreds more.

The babies that hatched from the lobster's eggs did not look like her. They were big-eyed and transparent, about the size of mosquitoes. Nor did they behave like lobsters. They rose and fell aimlessly near the surface and tried to eat everything in sight, including each other.

Lobster mothers carefully tend their eggs, but they never stay with their young after they hatch. Since lobsterlings are food for many sea creatures, not many survive without the mother's protection.

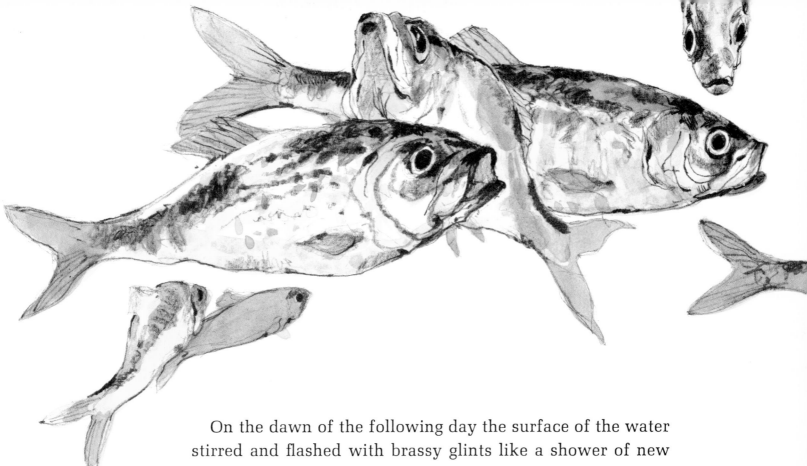

On the dawn of the following day the surface of the water stirred and flashed with brassy glints like a shower of new coins. A school of young herring were sifting the water with large gaping mouths to strain out any floating particles of food. The little lobsters, knowing no fear, bobbed innocently in their path.

The fish had already eaten half the babies when suddenly their orderly ranks broke in panic and confusion. Bluefish were attacking. As the blues tore through their school, the herring fled in all directions. The remaining lobsters drifted, unconscious of their narrow escape.

More of them were killed by other surface-feeding fish and sea birds. Wind and currents scattered the few that were left until one of them found herself alone, many miles from the place where she was born.

The lobsterling shed her shell four times in the next few weeks until the shell underneath looked just like her mother's, except in one special way. Instead of being greenish brown like other lobsters, she was, by an accident of nature that sometimes happens, a glorious blue.

Her eyes were on movable stems. She could see in all directions, but she could see only dimly. She had to rely on touch hairs all over her shell to feel vibrations in the water. And with her new lobster body came an instinct that told her to leave the light and hide from her enemies in the loose stones near the shore.

As a person's bones grow, his muscles and skin grow at the same time. But a lobster doesn't have bones. Its skeleton is the hard shell that covers all of its body. The shell cannot grow, and as the lobster's body grows inside, its shell becomes too tight.

Soon after the blue lobster began living on the bottom, she became restless and uneasy. She felt a great swelling from within, as though her shell would split apart . . . and then it did. It split right across the back and the lobster struggled free of it.

Underneath, a new shell had formed, but it was still soft and she was defenseless. She dared not look for food, even under the cover of darkness. She ate the old shell. It was full of calcium, which would help to harden the new shell. During the following week while the new shell was still soft, she was able to grow larger.

When winter came, she moved to deeper water and burrowed in the mud. Lobsters are cold-blooded, which means their bodies adjust to the outside temperature. When the temperature was low, she ate less, grew less and slowed to the point of appearing lifeless.

By her first birthday the next summer she had outgrown her shell ten times and was two inches long. She lay in ambush in the swaying eel grass, catching small crabs and fish that came within reach.

As the years passed, she grew in size and strength. She was able to hunt for shellfish, cracking them with her large crusher claw. Her delicate second pair of walking legs picked out the meat and brought it to her mouth where it was shredded.

She learned always to be alert to danger. Once, as she plowed for clams, the sand began to ripple. A flat, buried object broke away from the bottom. The lobster quickly backed under a rock as a large skate, who would have eaten her, swooped off like a kite dipping in the wind, steering with his long tail.

When she was six years old, she weighed about a pound, the size her mother had been when the blue lobster was born. Like her mother she hid by day in a rock crevice, only leaving at night when she was hungry.

One evening a fishy odor drifted into her fortress. Her sensitive antennae picked up the lovely smell. They waved back and forth.

The promise of a good meal lured her from the rocks. She set out in her blue shell armor, its touch hairs guiding her through the gloom like a cat's whiskers.

The odor of food grew stronger. The lobster's excited antennae waved faster, her mouth parts worked, her beady eyes stretched on their long stalks.

And then she saw the appetizing chunks of fish cradled in a wooden basket. She was just about to tiptoe up and eat when her touch hairs felt the warning vibrations of an intruder.

She whirled around, claws raised in defense. Another, larger lobster had come to dinner. The newcomer shoved the blue lobster to show his superior strength. Far beneath the path of the full moon across the water the two lobsters locked their great claws together. They strained and heaved with all their might. The struggle was slow and silent.

One of the strange things a lobster can do is detach its leg or claw at certain joints and grow one back again. But if a claw is torn off in the wrong place, the lobster might bleed to death. Since the blue lobster could not break the other's hold, she chose to detach her smaller claw and escape. She flipped her tail and shot backward, stirring up a screen of mud to cover her retreat. She scurried to the nearest shelter, an abandoned rubber boot.

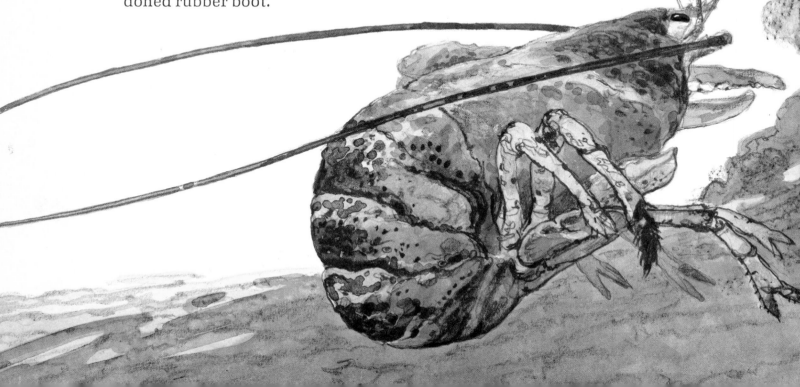

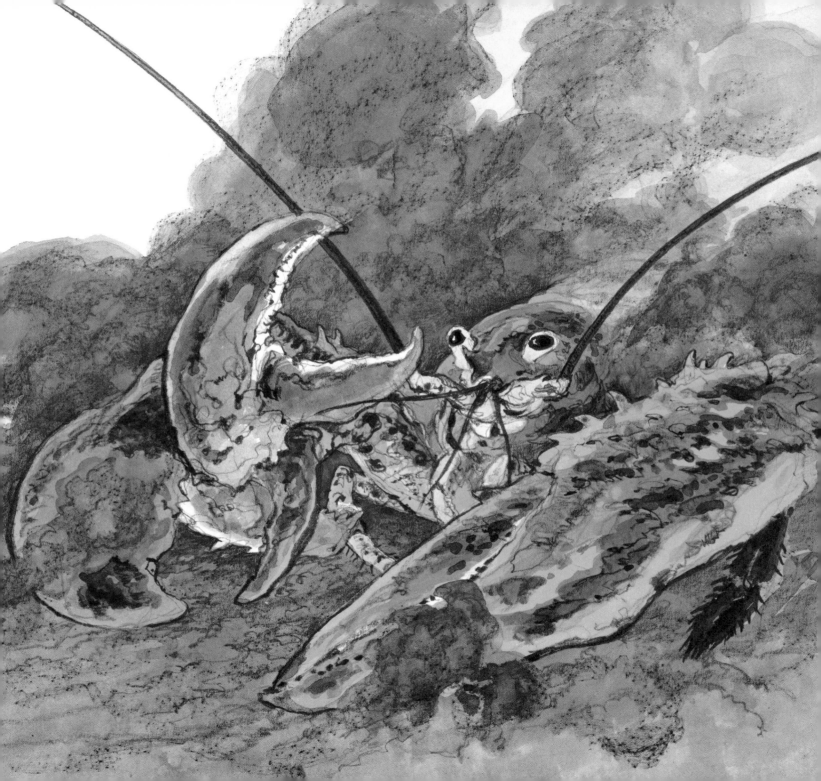

While the blue lobster cringed in the boot, the winner entered the basket to claim his prize. He ate with a great deal of energy, tearing and flinging the fish about, spitting up the undigestible parts.

But when the well-filled lobster tried to find his way out, he discovered the basket was, in fact, a trap.

When the morning light began to filter through the inky water, time for lobsters to return to the safety of their caves, both heard the sound of oars striking a boat. It was the sound of man, who is the lobster's worst enemy.

The lobster pot was raised slowly out of sight, its prisoner never seen again. The basket was lowered baited with fresh fish.

Peering from her hiding place in the boot, the lobster could see that the new bait had attracted another killer—a dogfish shark. His nose was the keenest of all the fish, and his teeth were especially designed to crush a lobster.

The shark began to circle slowly, then faster and faster. It made a sudden lunge and snapped at the trap. And then just as suddenly it swam off.

As adults, lobsters grow at a much slower rate and need to shed only once a year. The next time the blue lobster shed, her new shell had formed the bud of another claw.

While she was still soft, a male lobster was attracted to her. He began a dance on the tips of his walking legs. His long antennae waved back and forth, stroking her. Then he gently rolled her onto her back and deposited a fluid called sperm into a special pocket in her body. Eggs were already growing inside her, but without this sperm they would not develop into baby lobsters.

She carried the eggs for nine months until one day she again turned onto her back. The lobster forced out her eggs in a sticky mass and added the sperm that she had been saving in the special pocket all this time. The eggs stuck to her swim-merets for another nine months until they hatched.

Few of her children would live to shed their shells as many times as she or learn all she had about survival.

In the lobster's old age she was seldom seen outside her cavern under the deep water of the ocean. Her bright blue shell was now jeweled with crusts of barnacles and waved with seaweed streamers. She lived to the rare ripe old lobster age of twenty and grew to weigh almost thirty-five pounds.

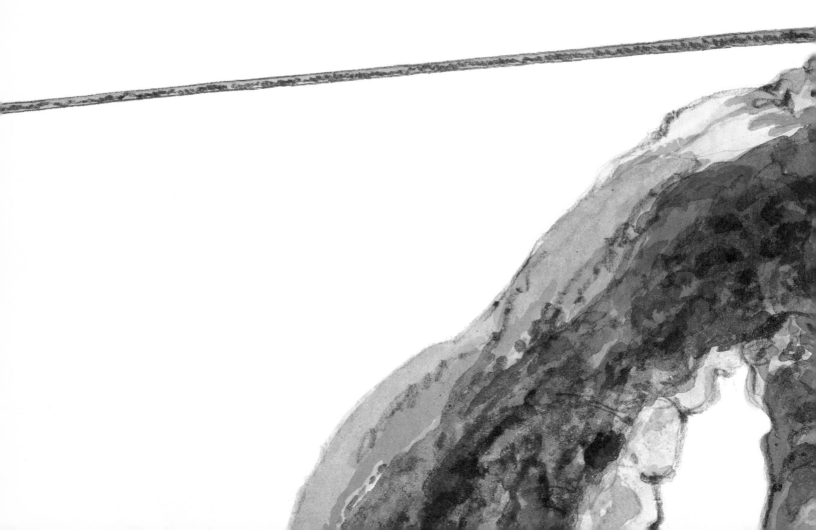

## About the Author and Artist

Carol and Donald Carrick have collaborated on a number of children's books on nature themes, including BEACH BIRD and A CLEARING IN THE FOREST. Donald Carrick is also the illustrator of BEAR MOUSE, selected for the 1974 Children's Book Showcase.

Carol Carrick was born in New York and was graduated from Hofstra University. She has worked as an art director for an advertising firm but now devotes her time to writing.

Donald Carrick was born and raised in Dearborn, Michigan. He studied art at Colorado Springs Fine Arts Center and at the Vienna Academy of Art in Austria. After traveling and painting in Europe, he settled in New York, where he has had several one-man shows.

The Carricks now live on Martha's Vineyard, Massachusetts, with their two sons. In researching THE BLUE LOBSTER, they have been able to draw on the resources of the State Lobster Hatchery and Research Station located on the island.